Editorial Board
编委会

Chief Consultant	总 顾 问	余 工	Yu Gong
Consultants	顾 问	余 敏	Yu Min
		周 晓 霖	Zhou Xiaolin
		罗 照 球	Luo Zhaoqiu
		余 佳 峻	Yu Jiajun
Editorial Board Director	编委会主任	郝 峻	Hao Jun
Editorial Board Members	编 委	徐 橘 Xu Ju	冯 华 忠 Feng Huazhong
		郭 晓 华 Guo Xiaohua	何 焱 生 He Yansheng
		刘 小 平 Liu Xiaoping	冷 祖 良 Leng Zuliang
		李 幼 群 Li Youqun	叶 武 建 Ye Wujian
Chief Editor	主 编	侯 江 林	Hou Jianglin
Text Editors	文 字 编 辑	北 雁	Bei Yan
		梁 海 册	Liang Haishan
Art Editor	美 术 编 辑	刘 晓 蒙	Liu Xiaomeng
Coordinator	统 筹	叶 玉 琴	Ye Yuqin

Stellar Thoughts

星思维

第七届"星艺杯"设计大赛获奖作品集

星艺装饰文化传媒中心 编著

暨南大学出版社
JINAN UNIVERSITY PRESS

中国·广州

图书在版编目（CIP）数据

星　思维：第七届"星艺杯"设计大赛获奖作品集 / 星艺装饰文化传媒中心编著. —广州：暨南大学出版社，2019.8
ISBN 978 - 7 - 5668 - 2686 - 2

Ⅰ.①星…　Ⅱ.①星…　Ⅲ.①建筑设计—作品集—中国—现代　Ⅳ.①TU206

中国版本图书馆CIP数据核字（2019）第164547号

星　思维：第七届"星艺杯"设计大赛获奖作品集
XING SIWEI: DIQIJIE "XINGYIBEI" SHEJI DASAI HUOJIANG ZUOPINJI
编著者：星艺装饰文化传媒中心
....

出 版 人：	徐义雄
策划编辑：	黄志波　杜小陆
责任编辑：	黄志波
责任校对：	王燕丽
责任印制：	汤慧君　周一丹

出版发行：暨南大学出版社（510630）
电　　话：总编室（8620）85221601
　　　　　营销部（8620）85225284　85228291　85228292（邮购）
传　　真：（8620）85221583（办公室）　85223774（营销部）
网　　址：http://www.jnupress.com
排　　版：广州良弓广告有限公司
印　　刷：深圳市新联美术印刷有限公司
开　　本：889mm×1194mm　1/12
印　　张：18.5
字　　数：170千
版　　次：2019年8月第1版
印　　次：2019年8月第1次
定　　价：198.00元

（暨大版图书如有印装质量问题，请与出版社总编室联系调换）

Create
Happiness
And
Deliver Joy

设计幸福　播种快乐

星 艺 装 饰 文 化 传 媒 中 心

1 住宅·工程实景作品

金奖 • 002 — 雅居乐剑桥郡

银奖 • 006 — 洞见
010 — 亚运城山海湾

铜奖 • 014 — 河北北戴河戴河林语
016 — 美的林城时代
018 — 合景天骏

优秀奖 • 022 — 金地天际
026 — 林间趣味
030 — 淮安半岛
032 — 万科·金域华府
036 — 珠江帝景楼王
038 — 中航城复式楼
042 — 乐湾国际
044 — 贵阳中铁国际城
046 — 人居紫云庭
048 — 山居秋暝

2 住宅·方案设计作品

金奖 • 052 — 天誉半岛

银奖 • 056 — 江川悦城
060 — 文山畅林苑公寓

铜奖 • 064 — 青浦周宅
068 — 创鸿水韵尚都
074 — 翠岛天成

优秀奖 • 076 — 保利中辰广场
080 — 腔调
084 — 三江尊园
088 — 汕头林宅
092 — 天地源御湾别墅
096 — 御龙首府复式楼
098 — 创新花园别墅
102 — 荣盛华府
104 — 中航城别墅
106 — 紫泥镇别墅

3 公共·工程实景作品

奖项	页码	作品
金奖	110	卓思中心
银奖	114	桂林香樟林别院
	118	宿静
铜奖	122	木空
	124	同和设计中心
	128	塞纳多皮肤中心
优秀奖	132	强声纺织
	134	桂林岩兰酒店
	138	"白房子"创客中心之 418 CUCOLORIS
	140	瓶子酒吧
	142	点心时代
	144	东方极韵
	146	李白啤酒馆
	148	苋泰科技
	150	天佑销售中心
	154	贺州星艺业务部

4 公共·方案设计作品

奖项	页码	作品
金奖	158	创举办公室
银奖	164	至和大厦顶层会所
	170	稻之源料理
铜奖	176	凤凰空间艺术馆
	180	森山亭
	184	航帆酒店
优秀奖	188	尚·灰
	190	诚信保险代理呼和浩特市总部
	192	索兰国际
	194	数控中心
	196	同心美术馆
	198	一朝一春（伊豆原）
	200	在水一方
	202	迈卡酒店
	204	聚和兴茶庄
	206	素菜馆

1

住宅·工程实景作品
Residence·Engineering Live-scene Works

雅居乐剑桥郡

Yajule Cambridgeshire

本案设计整体拒绝烦琐的材料堆砌，舍去多余的墙面造型，用明快的线条比例和质感对比表现简洁大方的空间设置。

The design of this scenario rejects the cumbersome material stacking, and eliminates the extra wall shape, expressing the simple and generous space setting with bright line proportion and texture contrast.

项目名称：雅居乐剑桥郡
项目设计：广东星艺装饰集团
项目地址：广东广州
设计师：刘永明

设计师用极简解构的手法把空间打散重组，在满足实用功能的基础上，让空间既有区域分割感，又能呼应整体。

没有刻意的造型设计，而是在结合实用需求的前提下，再去考虑线条比例和材质搭配。

空间的地面、天花、墙面都是独立的体块，更像是将几个体块组装起来的空间。餐桌与吧台的结构穿插，让空间的设计感更强，趣味性更浓。餐桌上方长条的灯具与餐桌相呼应，把空间进一步拉伸放大。

The designer uses the minimalist deconstruction method to break up and reorganize the space. On the basis of satisfying the practical functions, the space has both a sense of regional division and echoed as a whole.
There is no deliberate design, but on the premises of practical needs, then to consider the line proportion and material mix.
The ground, ceiling, and wall of the space are independent blocks, more like a space for assembling several blocks. The structure of the dining table and the bar is interspersed to make the design of the space stronger and more interesting. The long strip of lamps above the table echoes the table and further expands the space.

洞见
Super Insight

本案将垂直园林的理念融入住宅设计，交通空间交叠，如园林般的私密、亲切、居住性的尺度被张弛有度地塑造出来。运动与时间糅合为一种动态的现实，似垂直廊道穿行于院落之间。以朴素谦和的直纹白橡木饰面带入自然感，空间转折与明暗之间犹如园林，充满色彩、光影和温度，使空间体验变得更加微妙。

In this scenario, the concept of vertical gardens was incorporated into the residential design, and the traffic space was overlapped, forming the intimacy, and inhabitability of the garden. Movement and time are combined into a dynamic reality, like a vertical corridor between courtyards. With simple and straight white oak veneer, it brings a sense of nature. The space between the corner and the light and shade is like a garden, full of color, light and temperature, making the space experience more subtle.

项目名称：洞见
项目设计：广东星艺装饰集团
项目地址：广东广州
设计师：谭立予

1 住宅·工程实景作品
Residence · Engineering Live-scene Works

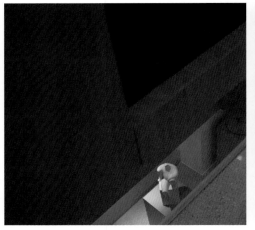

亚运城山海湾

Asian Games City Shanhaiwan

亚运城山海湾位于广州南部，风景独特的莲花山风景区的南麓，莲花山水道西岸，是 2010 年亚洲运动会的一项大型工程项目。

The Asian Games City Shanhaiwan, a large-scale project of the 2010 Asian Games, is located in the southern part of Guangzhou, with the unique scenery of the Lianhua Mountain Scenic Area in the south and the Lotus Mountain Waterway on the west bank.

项目名称：亚运城山海湾
项目设计：广东星艺装饰集团
项目地址：广东广州
设计师：吴明太

本案业主是位IT精英男，海归，习惯了西方的生活方式，偏爱白色。为了满足一家人的生活需求，设计师将两套复式合并成一个大复式。楼梯退居大门左侧角落，两套房子的客厅合二为一，变身拥有270度无敌视野的超大客厅。设计采用了纯净的黑白灰以及宁静的原木色，为了打破大面积运用白色的单调性，采用了黑色的天花线条以及彩色的装饰画、条纹抱枕来丰富视觉层次，还做了冷色与暖色交织的双色灯光设计。

The owner of this scenario is a male IT elite, overseas returnee, used to the Western way of life, preferring white. In order to meet the whole family's living needs, the designer merged the two sets into a large complex. The stairs is in the left corner of the main entrance, and the living rooms of the two houses are combined into one, transforming into an oversized living room with an invincible view of 270 degrees. The design uses pure black-and-white ash and quiet wood color. In order to break the monotony of white used in large areas, black ceiling lines and colored decorative paintings, striped pillows are used to enrich the visual level, and the cool and warm colors are interwoven to decorate the light.

室内的顶、墙、地六个面，完全不用纹样和图案装饰，只用线条、色块来区分点缀。线条利落简洁，除了简单的直线和直角外，沙发、茶几、桌子都是直线或简洁流畅的曲线条，造型简单，富含哲学意味，但不夸张。

饰品、灯具多采用错落有致的几何体形状。整体设计巧妙地运用块面造型和富有设计感的家具饰品，勾勒出明晰的空间层次，西厨吧台背靠沙发，起到了分割客厅与西餐厅的作用。老人房和小孩房增加了一些暖灰色，调和粉色调，提升空间的温馨指数。楼梯下，大门通道，隐藏式的收纳设计使整个空间看起来更加纯粹，每一物件都强调相对于人的机理功能学和贮物使用合理性。阳光铺满整个房间，浪漫惬意，优雅清新，生活的味道如此美好！

The top, wall and floor of the room are completely decorated without patterns, and only lines and color blocks are used to distinguish the embellishments. The lines are neat and succinct. In addition to simple straight lines and right angles, sofas and tea tables are straight lines or simple and smooth curves. The shape is simple, rich in philosophical meaning, but not exaggerated.

Jewelry and lamps often use a geometric shape that is patchwork. The overall design subtly uses the block shape and the design furniture to create a clear spatial level. The western kitchen bar is backed by the sofa, which plays the role of dividing the living room and the western restaurant. The old man's room and the child's room have added some warm gray, blending the color of the powder and enhancing the warmth index of the space. Under the stairs, the gates and the hidden storage design make the whole space look more pure, and each item emphasizes the rationality of the mechanism and storage of the objects relative to humans. The sunshine is shining the full room, romantic, elegant and fresh, and the taste of life is so beautiful!

河北北戴河戴河林语

Daihe Linyu in Hebei Beidaihe

每个时代都有属于当时的设计语言，每个细枝末节都蕴藏着最真实的生活状态。本案业主是一对年轻夫妇，喜欢中式优雅质朴的气质。设计师将中式元素与现代家居完美结合，环境、氛围皆让人有禅静如水的感觉，呈现出一种跨越年龄、包容的大气舒心之美。

Every era has its unique design that belongs to the time, and every detail embodies the most authentic state of life. The owner of this scenario is a young couple who likes the elegance and simplicity of Chinese style. The designer combines Chinese elements with modern homes. The environment and atmosphere give people a feeling of meditation and inner peace, showing a beautiful and comfortable atmosphere that spans age and tolerance.

项目名称：河北北戴河戴河林语
项目设计：广东星艺装饰集团
项目地址：河北秦皇岛
设计师：刘天亮

美的林城时代

Midea Lincheng Times

本案为洋房顶楼复式楼，开间不够开阔。设计师以"退"的姿态让空间形成多层次，从而形成"进"的感觉。这些符合东方人的思维习惯：在"退让"和"遮掩"中内敛地表现"层次"和"深度"。通过电视背景墙的大理石纹理体现山水写意的情调；深色的实木线条与白色的天花在暖白色的灯光照耀下，营造出东方精致细腻的宁静氛围。米色的墙面通过乌金木的黄色过渡与咖色的布面沙发搭配，给人简洁利落的感觉，恰恰符合化繁为简的新中式风格的精髓。

项目名称：美的林城时代
项目设计：广东星艺装饰集团
项目地址：贵州贵阳
设计师：田晓汐

This scenario is a duplex building on the top floor of the house. The opening space is not wide enough. The designer uses a "retreating" attitude to form a multi-level space, thus forming an "entering" feeling. These are in line with the inertia of the Orientals: in the "concession" and "covering" introverted "level" and "depth". The marble texture of the TV background wall reflects the romantic mood of the landscape; the dark solid wood lines and the white ceiling are illuminated by the warm white light, creating a delicate and peaceful atmosphere of the East. The beige wall is matched with the coffee-colored cloth sofa by the yellow transition of Zingana, sending out a simple and neat feeling, which is in line with the essence of the new Chinese style.

合景天骏

Hejing Tianjun

珠穆朗玛峰是喜马拉雅山脉的主峰，为世界第一高峰，巍峨宏伟，气势磅礴。攀登珠峰是最具挑战性的极限运动，总是有那么一种力量，让人们宁愿用生命作代价也要前仆后继地登上这并不适合人类生存的世界最高峰。因为这代表了一种梦想，一种勇于进取的精神和征服艰难困苦的毅力。

Mount Everest is the main peak of the Himalayas, the world's highest peak, grand and magnificent, making climbing it the most challenging extreme sport. There is always a kind of power that lures people into sacrificing their lives to conquer the highest peak in the world that is not suitable for human survival. Because this represents a dream, a spirit of courage and advancement and perseverance to overcome hardships.

项目名称：合景天骏
项目设计：广东星艺装饰集团
项目地址：广西南宁
设计师：南宁集成公司

　　本案的业主喜好运动，很有竞技精神，事业发展也有良好的上升空间，因此本案的设计以珠峰为主题以契合业主的精神品格。

　　公共空间整体简约的现代风格中还隐约透着些许轻奢的品格。4.8米层高的空间里，黑白灰的用色表达雪山的肃穆大气，添一抹蓝既是蓝天也如清新的空气让人感觉宁静。一组珠峰的主题油画对空间文化意境的提升起到了点睛的作用。

The owner of this scenario likes sports, has a competitive spirit, and has a good career development. Therefore, the design of this scenario is based on the theme of Mount Everest to suit the spirit of the owner.
The overall simplicity of the public space is also faintly revealing a little extravagant character. In the space of 4.8 meters high, the color of black and white gray expresses the solemn atmosphere of the snow mountain, adding a touch of blue like a blue sky and a fresh air makes people feel quiet. A set of the thematic oil paintings of Mount Everest has played a decisive role in the improvement of the spatial cultural conception.

金地天际
Golden Space

空间是一个载体，它承载了我们的生活与灵魂。设计师是空间设计的作者，笔尖引导着空间主人生活的走向，绘制着他们的灵魂高度。

Space is a carrier that carries our life and soul. The designer is the author of the space design, and the nib guides the direction of the space owners' life and draws the height of their soul.

项目名称：金地天际
项目设计：广东星艺装饰集团
项目地址：江苏常州
设计师：刘 峰

客厅中使用的材质较为丰富，大面积的木饰面以及大理石让空间显得更加高级，同时两者的搭配能够让空间温度趋于平衡。

The materials used in the living room are rich. The large wooden veneer and the marble make the space look more advanced, and the combination of the two can perfectly balance temperature of the space.

餐厅是一个具有生活仪式感的地方，设计师在这一部分的处理却显得十分俏皮，无规则的顶灯以及不同款式的座椅搭配，让一切变得有趣起来。

The restaurant is a place with a sense of life ritual. The designer's treatment is very playful in this part. The irregular ceiling lights and the different styles of seats make everything interesting.

林间趣味
Forest Fun

淡雅的空间背景，搭配橄榄绿，是暖意温存下的风度翩翩，是悠然肆意里的沁人心脾，源自大自然，牵引着源源不断的活力与生机，在自然的氛围里感受着浪漫优雅。

The elegant space background, with olive green, is the grace of warmth and relaxation. It can refresh people's heart, from nature, drawing a steady stream of vitality, and thus let people feel the romantic elegance in a natural atmosphere.

项目名称：林间趣味
项目设计：广东星艺装饰集团
项目地址：江苏苏州
设计师：李 娜

神秘而稀有的绿光散发着属于自然的魅力与独到的光泽。

The mysterious and rare green light exude the charm of nature and the unique luster.

蜂蜜色的金属装饰材质,增加了空间的质感,也收敛了原空间太过清雅的风格,让自然的效力恰到好处。

The honey-colored metal decorative material not only adds to the texture of the space, but also cools down the original space, so that the natural effect reaches a fine balance.

淮安半岛
Huaian Peninsula

项目名称：淮安半岛
项目设计：广东星艺装饰集团
项目地址：福建福州
设计师：刘小勇

　　本案设计为典雅的欧式风格，在造型上极其讲究，给人端庄典雅、高贵华丽的感觉，具有浓厚的文化气息。在家具选配上，采用宽大精美的家具，配以精致的雕刻，整体营造出一种华丽、高贵、温馨的感觉。

　　在配饰上，金黄色和棕色的配饰衬托出古典家具的高贵与优雅，赋予古典美感的窗帘和地毯、造型古朴的吊灯使整个空间看起来富含韵律感又大方典雅，柔和的浅色花艺为整个空间带来了柔美的气质，给人以开放、宽容的非凡气度，丝毫不显局促。

　　本案在色彩上以浅灰绿色系、黄色系为基础，搭配深棕色、金色等，表现出古典欧式风格的华贵气质。

The design of this scenario is of elegant European style, which is extremely particular in style, giving people a dignified, noble and gorgeous feeling, with a strong cultural atmosphere. In the selection of furniture, the use of large and exquisite furniture, accompanied by exquisite carving, creates a gorgeous, noble and warm feeling.

In the accessories, the golden and brown accessories highlight the elegance of classical furniture. The curtains and carpets with classical aesthetics and the quaint chandeliers make the space look rich and elegant. The soft, light-colored floral art brings a feminine temperament to the entire space, giving people an open, tolerant and extraordinary attitude.

The color of the scenario is based on light gray green and yellow, with dark brown and gold, showing the luxurious temperament of classical European style.

万科·金域华府
Vanke · Jinyu Huafu

设计应该认真考虑地球的有限资源使用问题，为保护地球的有限资源服务。所以本案设计更多的是去做减法设计，剔除一些过繁的元素与符号。

The design should carefully consider the limited use of the Earth's resources and serve the limited resources of the Earth. Therefore, the design of this scenario is more towards subtractive design, eliminating some of the excessive elements and symbols.

项目名称：万科·金域华府
项目设计：广东星艺装饰集团
项目地址：广东广州
设计师：胡 哲

1 住宅·工程实景作品
Residence · Engineering Live-scene Works

本案设计注重的是功能的设计与空间的形体比例。整个空间以素白色为基调，搭配木色系列的家具，营造一个自然舒适的住宅空间。简约不等于简单，每处的细节都是经过深思熟虑后得出的设计和思路，不是简单的"堆砌"和平淡的"摆放"。

The design of this scenario focuses on the functional design and the physical proportion of the space. The entire space is based on plain white, with wood-colored furniture to create a natural and comfortable residential space. The whole design is simple but not plain. The details of each place are the designs and ideas that have been deliberately drawn. They are not simply "stacking" and "flat".

珠江帝景楼王

Regal Riviera King Tower

本案地处广州繁华都市的制高点，站在窗边可以感受珠江新城川流不息的城市节奏，有一种闹中取静的闲适感。超大的空间、挑高的客厅给人一种可以顺畅呼吸的感觉，整个空间以浅色系列为主，材质考究的大理石，柔软的布艺，浅色的墙纸，都在彰显空间不急不躁的氛围。像一杯浓香型咖啡，不喝都能闻到一股久远的香味。整个空间注重居住者诉求，流畅的动线，柔和的灯光，淡雅的材质，无不透出主人公文雅平静的性格。

项目名称：珠江帝景楼王
项目设计：广东星艺装饰集团
项目地址：广东广州
设计师：帅伯尤

This scenario is located at the commanding heights of the prosperous city of Guangzhou. Standing by the window, you can feel the rhythm of the city of Zhujiang New Town, where you can find tranquility in a hustling surrounding. The large space and the high living room give people a feeling of smooth breathing. The whole space is mainly composed of light colors with the exquisite marble material, soft fabric and shallow wallpaper, all contributing to the cozy atmosphere. It is like a cup of black coffee with the smell of long-lasting fragrance. The whole space pays attention to the occupants' demands, smooth moving lines, soft lighting, and elegant materials, all of which reveal the calm and elegant character of the protagonist.

中航城复式楼
Zhonghangcheng Duplex

项目名称：中航城复式楼
项目设计：广东星艺装饰集团
项目地址：贵州贵阳
设计师：冷 娟

经典黑白灰塑造出现代质感，匠心独运营造玄妙意境。一步一景，灵动婉约，至境至美。

雾色的墙面用灯带强调，钢灰色与魅影黑则显出属于现代的高级质感。画框、插花、摆件，精巧别致的软装犹如画龙点睛，让空间更显灵动。

The classic black, white and grey creates a modern texture, and the ingenuity creates a mysterious artistic conception. One view per step, it is very graceful and poetic.
The foggy walls are highlighted with lights, and the steel grey and phantom blacks show a modern high-quality texture. Picture frames, flower arrangements, ornaments, the delicate soft decoration adds finishing touches, making the space more flexible.

客厅一侧设置饮茶区，将现代几何与传统中式元素相结合，木石和谐生趣，营造出别具一格的悠然意境。过道简约的吊顶线条，尽头装饰淡雅的水墨莲花，画面生动，在黑白灰的空间里更显婉约清丽。

On the side of the living room, a tea drinking area is set up, combining modern geometry with traditional Chinese elements. The wood and stone are harmonious coexisted, creating a unique and artistic conception. The simple ceiling line of the aisle is decorated with elegant ink lotus flowers at the end. The picture is vivid, and it is more beautiful in the black, white and grey space.

开放式餐厅厨房设计使得空间更加开阔。满吊顶筒灯与飞鸟艺术吊顶设计,结合墙面"富贵有鱼"金属装饰品,让空间灵动异常。落地窗设计采光极佳,在阳光的沐浴下,整个餐厅都沉浸在轻盈的画面之中,至境至美。

The design of open dining kitchen makes the space even wider. Full ceiling downlights and flying bird art ceiling design, combined with the wall's "rich and fish" metal decorations, make the space agile. The floor-to-ceiling windows are designed with excellent lighting. Under the sun bathing, the whole dining room is immersed in the light picture, making people feel the artistic atmosphere.

乐湾国际
Yue Wan International

本案是法式风格，设计的时候将天使白、佩斯利紫、金棕等色彩综合运用，展现出居室的高贵典雅，简单金边线条、适宜的装饰呈现出豪华浪漫的格调，整体营造华贵舒适的柔美风情。

This scenario is a French style. On designing, it uses the colors of Angel White, Paisley Purple and Gold Brown, showing the elegance of the living room. The simple golden lines and the appropriate decoration present a luxurious and romantic style, which creates a luxurious and comfortable feeling.

项目名称：乐湾国际
项目设计：广东星艺装饰集团
项目地址：贵州贵阳
设计师：姚 辉

贵阳中铁国际城
Guiyang China Railway International City

居家生活的质感不是靠华丽物质的堆砌,而是借由适宜的灯光、家居家饰以及经过缜密计算的生活动线,交织成凌驾于奢华之上的纯粹生活空间。简单的拆分、精准比例的拿捏,以轻透的建材语言诠释"少即是多"的设计理念。

The texture of home life is not based on the mass of luxurious materials, but through the appropriate lighting, home furnishings and carefully calculated living lines, interwoven into a pure living space above luxury. Simple splitting and precise proportioning are used to explain the design concept of "less is more".

项目名称:贵阳中铁国际城
项目设计:广东星艺装饰集团
项目地址:贵州贵阳
设计师:廖 颖

人居紫云庭
Ziyun Courtyard

本案业主向往简洁明快的家居环境。于是，本案设计中，有现代的简洁，也有远山木与养料灰经典色的搭配。从生活需求以及生活品位出发，对建筑结构进行大量的改造与扩建，将功能系统重新分区，给生活与未来带来更多的美好可能。

The owner of this scenario is looking forward to a simple and bright home environment. Therefore, in the design of this scenario, there is a modern simplicity, and also a mix of distant mountain wood and classic gray. Starting from the needs and the quality of life, the building structure will be extensively remodeled and expanded to re-partition the functional system, bringing more beautiful possibilities to life and the future.

项目名称：人居紫云庭
项目设计：广东星艺装饰集团
项目地址：四川成都
设计师：张 羽

山居秋暝

Autumn Dusk

自然之风让生活在喧嚣都市中的人们返璞归真，以自然之美来诠释业主的人格美和理想中的生活之美。本案设计中，有山有水，宛如远离世俗。如诗如画，随意洒脱，毫不着力。放下烦恼，无忧无虑，全身心享受着生活带来的宁静、淳朴之美。

The natural wind returns the people living in the noisy city to the true nature, and interprets the personal beauty of the owner and the ideal beauty of life with the beauty of nature. In the design of this scenario, there are mountains and waters, and it is far from the world. The whole design, casual and effortless, presents an artistic world. Let go of troubles, carefree, and enjoy the peace and simplicity of life.

项目名称：山居秋暝
项目设计：广东星艺装饰集团
项目地址：重庆万州
设计师：黄聪华、王 欢

2

住宅·方案设计作品
Residence·Scenario Design Works

天誉半岛

Tianyu Peninsula

本案设计师通过揣摩空间形态，打造出具有温度感的低奢空间。硬朗的线条，柔和温暖的色调，充满质感的家具，体现出空间的质感。设计师为打破传统的模式，试图在线条的规模和错位中平衡视觉感受，追求程序和自由的变化。

The designer of this scenario created a low-luxury space with a sense of temperature by trying to work on the space form. Tough lines, soft warm tones, and full of textured furniture reflect the texture of the space. In order to break the traditional model, the designer tries to balance the visual sensibility and pursue the change of procedure and freedom in the scale and dislocation of the line.

项目名称：天誉半岛
项目设计：广东星艺装饰集团
项目地址：广东广州
设计师：帅伯尤、蔡志登

2 住宅·方案设计作品
Residence · Scenario Design Works

江川悦城
Jiangchuan Yuecheng

项目名称：江川悦城
项目设计：广东星艺装饰集团
项目地址：广西南宁
设计师：凌立成

本案业主是一名建筑师，对生活品质要求极高，希望把空间设计得简单实用、温馨舒适。

宽大的窗户设计令客厅更加明亮，显得空间更加干净利落。在家具选用上，大量运用收纳柜放置书箱、美酒、独特的装饰品等，让客厅和餐厅看起来非常有层次感，既有精致主义的质感又极富生活气息。卧室延续了简约的舒适风格，去除多余的摆饰，大面积木饰面墙壁搭配浅灰肌理的背景墙，没有一丝多余的色彩，简练而静谧。书房收纳柜的设计将空间利用到极致，宽大的落地窗设计更是通透明亮，无论是闲暇时的浅读还是在家工作，都能让业主有个赏心悦目的环境。

The owner of this scenario is an architect who has extremely high requirements for quality of life and hopes to design the space to be simple, practical and comfortable.

The large window design makes the living room brighter and the space cleaner. In the choice of furniture, a large number of storage cabinets, wine, unique decorations, etc., make the living room and dining space look very layered, both exquisite and rich in life. The bedroom extends over the simple and comfortable style, removing the extra ornaments, the large wooden veneer wall with the light gray texture of the wall, without a trace of extra color, concise and quiet. The design of the study cabinet makes the full use of the space, and the large floor-to-ceiling window design is transparent. The owner can have a pleasing environment to do some shallow reading or working.

2 住宅·方案设计作品
Residence · Scenario Design Works

文山畅林苑公寓

Wenshan Changlinyuan Apartment

本案以高级灰为主调，精致的家具、利落的线条以及木材、金属、玻璃和人造材料，都凸显出空间的时尚简约范儿。没有花哨的造型，也没有浓重的色彩，仅以简单而不失质感的空间设计，就给人带来一种宁静高贵的视觉感受。整个设计摒弃了烦琐浮华，以一种简约优雅的设计语言，结合对材质的精彩运用，极力呈现出低调、奢华的空间力量。

项目名称：文山畅林苑公寓
项目设计：广东星艺装饰集团
项目地址：云南文山
设计师：熊卫星

The scenario is dominated by high-grade gray, exquisite furniture, neat lines and wood, metal, glass and man-made materials, which all highlight the space's sleek minimalist style. There is no fancy shape, no strong color, and the design of the space is simple but not loosing feel, giving a quiet and noble visual experience. The whole design abandons the cumbersome and flamboyant, with a simple and elegant design language, combined with the wonderful use of materials, striving to present a low-key, luxurious space power.

以大面积的高级灰作为铺陈，是现代轻奢风格的最佳演绎。它仿佛拥有一种直撩心底的魔力，轻轻松松就能将沉稳、宁静的一面最大化地彰显。随处可见的隐藏式灯带，在强化空间奢华气质之余，带来一种高品质的空间享受。开放式的空间布局，让视觉更开阔、更明朗，再结合巧妙的灯光点缀和对比色的应用，突出层次感，给空间带来灵动多变的视觉效果。

With a large area of high-grade gray as the pavilion, it is the best interpretation of modern luxury style. It seems to have a magical power that is straightforward, and it can easily show the calm and quiet side. The hidden light strips everywhere can enhance the luxurious temperament of the space and bring a high quality space.The open space layout makes the visual more open and clear, combined with clever lighting and contrasting applications, the layering is outstanding, giving the space a dynamic and varied visual effect.

青浦周宅

Qingpu Zhouzhai

项目名称：青浦周宅
项目设计：广东星艺装饰集团
项目地址：上海青浦
设计师：范建国、许 舰

心之所向，便为生活之所望。
Heart's yearnings are the hope of the life.

本案位于林荫河畔，自在且悠闲。整个空间以素雅的黑白灰色调为主，辅以温润的原木色调作为点缀。简练的直线条体现了业主对简单生活的追求和态度。工艺精湛的金属灯具和舒适的灰调子麻质沙发相得益彰，表现出业主对生活的细致要求。石材的选择遵循整体空间氛围规则，纹理低调柔和，不张扬夺目。冷暖灯光的配合加强了空间的层次感，突出了雕塑式的空间形体构造。

This scenario is located on the bank of the river, free and leisurely. The entire space is dominated by elegant black and white gray tones, complemented by warm wood tones. The concise line shows the owner's pursuit and attitude towards simple life. Crafted metal lamps and comfortable gray-toned linen sofas complement each other, showing the owner's meticulous requirements for life. The choice of stone follows the rules of the overall space atmosphere, combined with soft texture, presenting an atmosphere without any publicity. The combination of cold and warm lighting enhances the layering of the space and highlights the sculptural spatial form.

创鸿水韵尚都

Chuanghong Fashion City by Waterside

设计师巧妙地运用了建筑与光线之间的趣味关系,让人漫步其中,可以感受到室内和室外之间情感的交流互动。
一层设置为公共空间区,三面被阳光和院子包围,室内外光影交错,在室内也能随时随地接受阳光的温暖拥抱。

项目名称:创鸿水韵尚都
项目设计:广东星艺装饰集团
项目地址:广东佛山
设计师:姚国健

The designer skillfully uses the interesting relationship between architecture and light, allowing people to walk around and feel the emotional interaction between indoors and outdoors.
The first floor is set up as a public space area, surrounded by sunlight and yards on three sides. The indoor and outdoor lights are interlaced, and the warm embrace of the sun can be received anytime and anywhere.

男业主是一名著名的服装品牌创始人和总设计师。为满足业主的需求，打造一个有质感和空间感的家，设计选用了白色亚光的大理石墙身，连接天花木饰面和地面的珍珠灰，表现一个简约、轻奢，睿智、轻松的空间。

二层呈中轴对称分开，左边留给未来的小孩居住，右边留给父母居住。

三层整层设计为主人房，左边以工作、休闲、储蓄为主，右边则是私隐的居住空间，功能齐全，动静分区，居住者可在此享受生活的乐趣。

The male owner is the founder and chief designer of a famous clothing brand. In order to meet the needs of the owners, the designer creates a home with a sense of texture and space. The design uses a white matt marble wall, connected to the ceiling wood and the ground pearl ash, showing a simple, mild luxury, wise and comfortable space.

The second floor is symmetrically separated by the central axis. The left side is reserved for future children and the right side is reserved for parents.

The owner's room is on the third floor. The left side is designed as a working area, accompanied by space of leisure and savings. The right side is the private living space. It is fully functional, dynamic and quiet, enabling the owner to enjoy the fun of life.

负二层主要是对外互动空间,设计师在空间材质上采用了温润的木色作为点缀,微妙的暖色给空间带来了温暖。采光井神秘的光线吸引着前往的脚步,休闲区开阔的空间给人带来了舒适感。

The negative second layer is mainly the external interaction space. The designer uses the warm wood color as the embellishment on the space material, and the subtle warm color brings warmth to the space. The mysterious light of the light well attracts the passersby, and the open space of the leisure area brings a sense of comfortable.

收藏展示区是业主个人的特殊要求,作为一名优秀的服装品牌创始人和总设计师,这里彰显着他独特的品位和审美。过道直接打通,连接南北采光井,让整体空间鲜活起来。

The collection-exhibition area is a special requirement of the owner. As an excellent clothing brand founder and chief designer, it highlights his unique taste and aesthetic. The aisle directly connects the north and south light wells to make the whole space alive.

翠岛天成
Cuidao Tiancheng

项目名称：翠岛天成
项目设计：广东星艺装饰集团
项目地址：河北秦皇岛
设计师：夏文彬、崔 健

本案注重线条的提炼和运用，把不同物体富有表现力的轮廓加以突出和强调，借以再现准确、鲜明的视觉形象，以丰富的线条表现形式来表现家居的质感、空间感和情感。不拘泥于对华丽外表的追求，而专注于宁静深邃的内心世界，从取材入手，强调时尚感。只要你身处其中，就能体会深远、惬意的感觉。

This scenario focuses on the refinement and application of lines, highlighting and emphasizing the expressive contours of different objects, in order to reproduce accurate and vivid visual images, and express the texture, space and emotion of the home with rich lines of expression. Not confined by the pursuit of gorgeous appearance, but focus on the quiet and deep inner world, starting from the material, emphasizing fashion sense. As long as you are in it, you can feel the profound and satisfying feeling.

保利中辰广场
Polly Sino Star Plaza

本案空间简洁，色调以黑白灰为主，整体给人一种宁静的感觉。在喧嚣的大城市，业主回到这个宁静的家，拿一份报纸或一本书，慢慢平复心情，静静享受生活。

项目名称：保利中辰广场
项目设计：广东星艺装饰集团
项目地址：广东广州
设计师：吴 佳

The space of this scenario is simple, with mainly black and white tones, and the whole gives a feeling of tranquility. In the hustle and bustle of the big city, on returning to this restful home, the owner could calm down his heart and appreciate the life while reading a newspaper or a book.

本案以浅色设计为主，材质方面选用自然的元素，如柔软质朴的沙麻布制品。其所呈现的现代风格以简约著称，具有浓郁的后现代主义特色，注重流畅的线条设计，代表了一种时尚，回归自然，崇尚原木韵味，外加现代、实用、精美的艺术设计风格。

This scenario is based on light-colored design, and mainly uses natural elements such as soft and rustic sand linen products. Modern style is famous for its simplicity, with rich post-modernism characteristics, focusing on smooth line design, representing a fashion, returning to nature, advocating the charm of the wood as well as modern, practical and exquisite art design style.

腔调
Accent

不追求外在，把舒适体验放在首位，极致追求功能与艺术的完美结合。

去繁求简，恰恰是快乐舒适的秘诀。拿起该拿的，放下该放的……

The design of this scenario puts the comfort experience in the first place instead of focusing on the appearance, thus, to pursue the perfect combination of function and art. It's the secret of happiness and comfort to seeking simplicity, in pursuit of the real needs.

项目名称：腔调
项目设计：广东星艺装饰集团
项目地址：广东潮州
设计师：徐 潇

2 住宅·方案设计作品
Residence · Scenario Design Works

不让俗世里多余的事物，来侵占我们的时间与经历。
将最纯粹的自我和情感释放其中，眼睛看到的，心灵感受到的，一定是一个极其沉静而本真的空间。

Do not let the excessive things in the world to encroach on our time and experience.
The purest self and emotion are released. Therefore, an extremely quiet and authentic space in the owener's eyes and heart must be presented.

三江尊园
Sanjiang Garden

项目名称：三江尊园
项目设计：广东星艺装饰集团
项目地址：内蒙古包头
设计师：梅 刚

　　本案业主是一位事业有成的精英女士，因为工作关系经常到世界各地出差，忙碌的工作和富裕的经济条件让业主对自己的生活方式有了新的追求：有文化感与高雅感，但不能缺乏自在感与情调感。设计师和业主经过多次沟通，最后确定装修风格为美式。
　　为满足业主的需求，设计师在整体的装修设计上，主要选用浅淡的暖色调和白色，用简洁的线条装饰，摒弃了过多的烦琐与奢华，既显得自然淡雅又凸显了良好的实用性。家具的选择上也采用了简化的线条、粗犷的体积，家具的颜色设计也相对淡雅，与整体色调相呼应。在材质的运用上多采用天然木、石材等元素，并且在后期的细节搭配中，融入各种自然的装饰元素，与美式家具贵气的轮廓相得益彰。再加上巧于设置室内绿化的精心点缀，没有太多刻意的雕琢与约束，营造一个浪漫的氛围，充分展示出主人追求的一种闲适、淡雅的自然生活方式。

The owner of this scenario is an elite woman with a successful career. She often travels around the world for work. Busy work and affluent economic conditions have made the owner have a new pursuit of her own lifestyle: a sense of culture and elegance as well as a sense of freedom and romanticism. The designer and the owner have communicated many times and finally determined the American style.
In order to meet the needs of the owner, the designer mainly uses light warm colors and white in the overall decoration design. It is decorated with simple lines, and has abandoned cumbersomeness and luxury, thus highlighting both nature and elegance as well as practicality. The choice of furniture also uses simplified lines and rough volume, and the color of the furniture is relatively mild, echoing the overall tone. Natural wood, stone and other elements are used, and in the later details, various natural decorative elements are incorporated, which complements the extravagant contours of American furniture. Coupled with the careful decoration of indoor greening, there is not much deliberate carving and restraint, creating a romantic atmosphere, fully demonstrating a leisurely and elegant natural lifestyle pursued by the owner.

汕头林宅
Shantou Linzhai

浮躁的生活需要一股清风，带来凉意，吹走喧嚣。

本案业主是一个四口之家，设计师认为构成家的重要元素是阳光。通过对原有户型的分析，设计师把一层的所有间隔墙体都打掉，将餐厨一体化，增加空间的对流与采光，同时增加了不同功能区的联系与互动。

The impetuous life needs a breeze, bringing coolness and blowing away bustle.
The owner of this scenario is a family of four, and the designer believes that the important element that constitutes a home is the sunshine. Through the analysis of the original apartment type, the designer knocks out all the partition walls of the first floor, integrates the kitchen and the dining area, thus increasing the convection and lighting of the space as well as the connection and interaction of different functional areas.

项目名称：汕头林宅
项目设计：广东星艺装饰集团
项目地址：广东汕头
设计师：林朝杰

把活动区放在一层，二层主要作为睡卧区，中空位的保留，楼梯造景与二层的落地窗，让两层空间有个很好的连接。
本案材质上采用灰色的水磨石、橡木饰面，以白色和木色为主色调，打造一个放松、慵懒、舒适的家。

Locating the moving area on the first floor, the second floor is the sleeping area, the retaining of the hollow space and the window between the staircase landscaping and the floor-to-ceiling of the second floor connect the two floors perfectly.
The material of this scenario is made of gray terrazzo and oak finishes, with white and wood tones to create a relaxed, lazy and comfortable home.

天地源御湾别墅

Tande Royal Bay Villa

生活品质的提升促使我们去追寻更深层的享受——舒适、优雅的生活态度，同时不失品位和高贵。

新东方主义，传统中式与现代元素巧妙兼容，中国传统文化意义在当代背景下重新演绎，将现代元素与传统元素结合在一起，以现代人的审美需求来打造富有传统韵味的空间。

The improvement of the quality of life has prompted us to pursue a deeper enjoyment—a comfortable and elegant attitude to life, without sacrificing taste and nobility.
Neo-orientalism embodies traditional Chinese and modern elements. The meaning of Chinese traditional culture is re-interpreted in the contemporary context, combining modern elements with traditional elements to create a space full of traditional charm with the aesthetic needs of modern people.

项目名称：天地源御湾别墅
项目设计：广东星艺装饰集团
项目地址：广东惠州
设计师：黄志平

御龙首府复式楼
Yulong Shoufu Duplex

项目名称：御龙首府复式楼
项目设计：广东星艺装饰集团
项目地址：福建龙岩
设计师：张丙华

本案采用现代风格设计，外形简洁、功能性强，强调室内空间形态和物件的单一性、抽象性。

现代简约，顾名思义，就是让所有的细节看上去都是非常简洁的。极简的装修风格能让空间看上去非常简洁、大气。装饰的部位要少，但是在颜色和布局上，在装修材料的选择搭配上需要费很大的劲儿，这是一种境界，不是普通设计师能够设计出来的。

目前，现代简约的装修风格更迎合年轻人的喜好。忙碌的都市生活让人身心疲惫，下班后，人们需要寻找一个安静、舒适的地方。借此，打造一个简约、宁静的生活空间，成了现代生活中的大需求，人们可以在这样的氛围中消除工作的疲惫，忘却都市的喧嚣。

This scenario is designed in a modern style with a simple shape and strong functionality, emphasizing the singularity and abstraction of the interior space form and objects.
Modern simplicity, as the name implies, makes all the details look very simple. The simplicity style of the decoration can make the space look concise and generous. There are few decorative parts, but in terms of color and layout, it takes a lot of effort to choose the matching materials. This is a realm that cannot be designed by ordinary designers.
At present, the modern and simple decoration style caters to the preferences of young people. Busy urban life makes people physically and mentally exhausted. After work, people need to find a quiet and comfortable place. In this way, creating a simple and quiet living space has become a big demand in modern life. People can eliminate the fatigue of work and forget the hustle and bustle of the city in such an atmosphere.

创新花园别墅

Chuangxin Garden Villa

现代简约风格就是简洁而有品位，这种品位体现在设计上的细节把握。简约主义风格的特色是将设计的元素、色彩、照明、原材料简化到最少，但对色彩、材料的质感要求很高。因此，简约的空间设计通常非常含蓄，往往能达到以少胜多、以简胜繁的效果。艺术创作宜简不宜繁，宜藏不宜露。

The modern minimalist style is simple and tasteful, and this taste is reflected in the details of the design. It is characterized by the simplification of the design elements, colors, lighting, and raw materials, but the demand of the colors and materials is very high. Therefore, the simple space design is usually very subtle, and often achieves the effect of doing more with less. Art creation should be concise rather than complicated and should be hidden but not exposed.

项目名称：创新花园别墅
项目设计：广东星艺装饰集团
项目地址：江苏常州
设计师：刘 峰

2 住宅·方案设计作品
Residence · Scenario Design Works

荣盛华府
Rongsheng Huafu

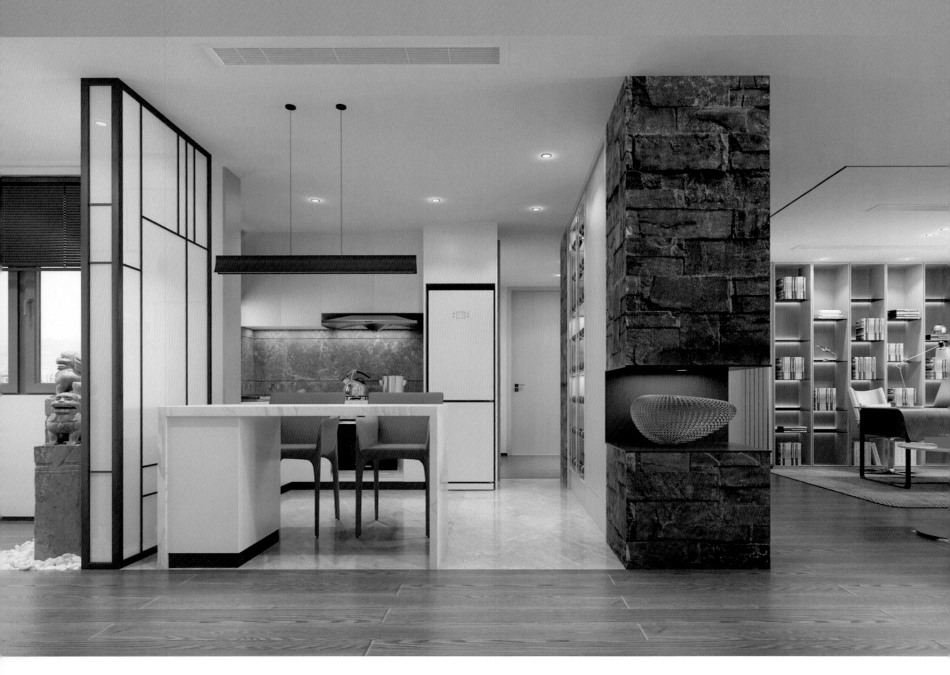

三房改一房的设计，一切以舒适与合理为原则，同时也要表现业主的个人品位。本案将所有公共空间打通，提高通透感。厨房和走道的打通尤为特别，配合文化石的使用，使酒柜那面墙体成为整个空间的视觉中心。书房采用了轨道门的隔断，敞开和关闭都各有一番天地。根据业主的需要，打造出理想中的宁静假日空间。

The design idea of converting three bedrooms to one bedroom is based on the principle of comfort and rationality, and also reflects the personal taste of the owners. This scenario will open up all public spaces and improve the sense of transparency. The opening of the kitchen and the walkway is especially special. With the use of cultural stone, the wall of the wine cabinet becomes the visual center of the whole space. The study has a partition of the track door, and there is a world of openness and closure. According to the needs of the owners, an ideal tranquil holiday space is created.

项目名称：荣盛华府
项目设计：广东星艺装饰集团
项目地址：安徽蚌埠
设计师：边诗琪

中航城别墅
Zhonghangcheng Villa

项目名称：中航城别墅
项目设计：广东星艺装饰集团
项目地址：贵州贵阳
设计师：李玉明、杨坚业

一园一院，依山而居。中航城依地而建，一眼望去，有"一同山居"的壮观感受。用建筑的语言来解读空间，以澄净、空渺的人文东方设计作为精神贯彻全案，这是设计师的初心。

Zhonghangcheng is built on the ground, and at first glance, there is a spectacular feeling of "living together with mountains" reaching the effect of "one garden one yard". It is the designer's initial intention to use the building language to interpret the space and carry out the whole scenario with a clean and empty humanistic oriental design.

紫泥镇别墅
Zini Town Villa

项目名称：紫泥镇别墅
项目设计：广东星艺装饰集团
项目地址：福建漳州
设计师：张 丹

本案使用极简风格设计，用理性的手法表现感性的生活。设计中简洁硬朗的线条造型搭配深色元素，简约而不失奢华。

设计采用深色木纹饰面板搭配灰色石材纹理地砖，凸显高级、时尚的氛围，使空间既张扬又不失低调，简洁又不失奢华，给人一种神秘的感觉。

卧室采用无主灯设计，巧妙运用筒灯及灯带照明，烘托出简洁时尚的氛围。深色木地板巧妙搭配木纹饰面板，提升了整个空间的格调，使其简洁又不失温馨。

This scenario uses a minimalist style design to express a sensual life in a rational way. The simple and tough lines in the design are matched with dark elements, simple yet luxurious. The design uses a dark wood veneer with gray stone texture tiles to highlight the high-end, stylish atmosphere, making the space both unassuming and understated, simple yet luxurious, sending out a mysterious feeling.
The bedroom is designed with no main lights, and the subtle spot lights and strip lights are used to create a simple and stylish atmosphere. The dark wood floors are cleverly matched with wood-grained panels to enhance the style of the space, making it simple yet warm.

3

公共·工程实景作品
Public·Engineering Live-scene Works

卓思中心
Zhuosi Center

本案为办公空间所属的景观院落。整个院落由开放式回廊串联起来，既是界面也是连接体，主视角由院内的风景和游历体验展开。里面隔栅恰当的密度带来半透明感，成为自然光的过滤器，延滞了视觉向内部的匆促穿透，让不同时段的日光更加柔和，使得多重光影交错。室内与室外的传统界限在这里被重组为观看与被观看的情景交互。

This scenario is the landscape courtyard to which the office space belongs. The entire courtyard is connected by an open corridor, which functions as both an interface and a connector. The landscape and travel experience are spread out with the view in the courtyard. The proper density of the inner grille brings a translucent feel, which becomes a natural light filter, delaying the rushing of the visual to the interior, making the daylight in different periods softer, making multiple light and shadow interlaced. The traditional boundaries between indoors and outdoors are reorganized here to interact with the scene being viewed.

项目名称：卓思中心
项目设计：广东星艺装饰集团
项目地址：广东广州
设计师：谭立予、郭晓华

3 公共・工程实景作品
Public · Engineering Live-scene Works

桂林香樟林别院

Xiangzhanglin Hotel, Guilin

本案位于桂林靖江王城边上，共有两层，一层为公区，二层为客房区。一层公区在有限的空间里通过动线的巧妙设置和水平高度的变化，充满了曲折回绕的乐趣。洁净的麦秆泥墙上锈铁的圆形院标和虚实结合的竹条栅大圆，构成了充满东方禅学意境的画面，也奠定了整个民宿的基调。

This scenario is located on the edge of Jingjiang Wangcheng in Guilin. There are two floors, the first is the public area and the second the guest room area. The first floor of the public area is filled with the joy of twists and turns through the clever setting of the moving lines and the change of the level in a limited space. The rounded yard of rusted iron on the wall of clean wheat straw mud and the large circle of the bamboo strips combined with the virtual and the real form constitute a picture full of oriental Zen, and also set the tone of the whole house.

项目名称：桂林香樟林别院
项目设计：广东星艺装饰集团
项目地址：广西桂林
设计师：范建国、许 舰

公区入口特地设置了换鞋柜，客人通过换鞋仪式将一切烦恼在进入别院后都抛之脑后。小巧的公区既有原木大茶桌，也有下沉式社交投影区，各取所需，自得其乐。在材料的选择上，全部使用老榆木原木板材，所有家具使用卯榫结构处理，充满了自然质朴的气息。麦秆藻泥和多孔火山岩板在温暖的灯光下与原木的温润共同营造出明亮、舒适、暖心的环境氛围。

The entrance to the public area has specially set up a shoe cabinet, and the guests are expected to put it all behind on entering the partial courtyard after changing shoes. The small public area has both a large tea table and a sunken social projection area, and thus the guests could get what they need and enjoy themselves. In the choice of materials, the old elm logs are used for all the areas, and all the furniture is treated with a birch structure, which creates a natural and rustic atmosphere. The stalked algae mud and porous volcanic slabs together with the warmth of the logs create a light, comfortable and warm environment.

二层客房区第一期开放7间房，以中国的7个节气命名。分别是：春分（思念）、谷雨（细腻）、夏至（旅行）、白露（芬芳）、小满（治愈）、立秋（守护）、冬至（温暖）。客房的榆木家具散发着岁月的味道。软棕乳胶床垫、流线型的大浴缸、大自然香甜的精油香薰都体现了主人对品质生活的追求和热爱。不同主题的客房散发着不同的独特气质：春分的透光帷幔、圆形浴缸、粉色的花卉摆件表现了浪漫的少女情怀；谷雨的原木格栅移门、布满手造草席的坡型屋顶、纯美的莲花油画和布艺抱枕，在质朴中散发着文人的清高气息；立秋的地台矮榻、趣味吊床特别适合有孩子的家庭入住……不同的选择带来不同的体验，让入住的客人褪去一天的劳顿，在城市中享受一片寂静。

Seven rooms is opened on the second floor room area and is named after China's seven solar terms, respectively, Chunfen(Spring equinox) (retrospection), Guyu(Grain Rain) (delicacy), Xiazhi(summer solstice)(traveling), Bailu(white dew)(fragrance), Xiaoman(lesser fullness)(healing), Liqiu(beginning of the autumn) (guardian), Dongzhi(winter solstice)(warm). The elm furniture of the room lingers with the taste of the years. The soft brown latex mattress, the streamlined large bathtub, and the natural sweet oil aroma all reflect the owner's pursuit and love of quality life. Different themes of the room have different unique temperament. The Spring equinox room's translucent enamel, round bathtub and pink floral ornaments represent the romantic girl's feelings; the log grille door of the Grain rain room, the sloping roof covered with hand-made straw mats, the pure lotus oil painting and the fabric pillow, exudes the literati's high spirit in the simplicity; the terraces and fun hammocks of the Liqiu room are especially suitable for families with children... Different choices bring different experiences, which enables the guests stay away from the day's work and enjoy the tranquilness in the city.

宿静
Sujing Hotel

本案由早期的老厂房改建而成，设计师没有对其做太多的改变，而是让"历史"延续并赋予它新的生命。室内拥有22个房间，由"趣味""民俗""自然"三个主题组成。

This scenario was transformed from an old factory building. The designer did not make too many changes to it, but let "history" continue and animates it. There are 22 rooms, which consist of three themes: "fun", "folk" and "nature".

项目名称：宿静
项目设计：广东星艺装饰集团
项目地址：上海嘉定
设计师：黄　剑

3 公共·工程实景作品
Public·Engineering Live-scene Works

木空
Mukong Hotel

本案项目取名为"木空"，是因为整个空间由5个极具趣味性的大盒子组成，设计师赋予每个盒子不一样的功能和使命。通过空间内不同程度的地面抬升和下沉、视线的隔断和重组，整个空间更具层次感。本案设计大量使用木材，秉承环保的可持续生态回圈，在设计之初植入绿色生态的概念。

The project was named "Mukong"(Wooden boxes) because the entire space consisted of five very interesting big boxes, and the designer gave each box a different function and mission. Through the different degrees of ground uplift and sinking in the space, the separation and reorganization of the line of sight, the whole space is more layered. This scenario uses a large amount of wood, adhering to the environmentally sustainable ecological cycle, and implanting the concept of green ecology at the beginning of the design.

项目名称：木空
项目设计：广东星艺装饰集团
项目地址：广东广州
设计师：吴家春

同和设计中心
Tonghe Design Center

本案位于广州市天河区与白云区交界处，地理位置优越，交通便利，附近有地铁站、公交车站和大型商场，人流量较大。原址是一个西餐厅，设计师在设计初期除了考虑正常的办公洽谈功能以外，最多的是考虑如何做到最大限度的节能环保，利用原有的材料，循环再用。

This scenario is located at the junction of Tianhe District and Baiyun District in Guangzhou, with the superior geographical position and the convenient transportation. There are subway stations, bus stations and large shopping malls nearby,so the traffic is busy. The original site is a western restaurant. In addition to considering the negotiation function of the office, the designer considers how to use and recycle the original materials thus achieving the maximum energy saving and environmental protection.

项目名称：同和设计中心
项目设计：广东星艺装饰集团
项目地址：广东广州
设计师：侯立原

3 公共·工程实景作品
Public·Engineering Live-scene Works

经过考虑，设计师把原有的地面层保留，将灯具、椅子、桌子加以改造利用。另外，在整体设计上，大面积利用颜色体块做功能区域的划分，室内和室外两个区域环境利用形体的穿插方式使其融为一体，把空间进一步扩大。

After consideration, the designer retained the original ground floor and renovated the lamps, chairs and tables for use. In addition, in the overall design, the large area is divided into functional areas by using color blocks, and the indoor and outdoor areas are integrated into the environment by using the weaving manner of the form to further expand the space.

塞纳多皮肤中心
XANADO Skincare Center

本案中明亮、轻奢风格的大厅给每个进门的客人一种眼前一亮的感觉。飘逸的白色纱帘、金色的轻奢家具都让人感觉自由随意。高大的标志树立在大厅中间，与四周弧形的墙面相呼应，既打破呆板的格局，也能瞬间吸引来客的眼球。

The bright, mild-luxury style of the hall in this scenario could inspire each guest on entering the gate. Flowing white gauze and gold-colored luxury furniture make people feel relaxed and free. The tall logo stands in the middle of the hall, echoing the curved walls around it, breaking the rigid pattern and instantly attracting visitors' attention.

项目名称：塞纳多皮肤中心
项目设计：广东星艺装饰集团
项目地址：湖南株洲
设计师：肖 冰

强声纺织
Giantsun Textile Company

项目名称：强声纺织
项目设计：广东星艺装饰集团
项目地址：江苏常州
设计师：王少兵

本案是一套老厂区办公室翻新设计，为了保留原厂区的特色，使其与其他楼层办公区氛围融为一体而挑选了LFOT工业风。设计意在打破沉闷古板、千篇一律的固有办公模式，创建一个热情且具有特色的办公环境。跨入科技感应门，设计师巧妙地运用不规则几何造型，再涂刷明黄、果绿色墙面，让整个空间充满了清新活力。以蜂巢的六边形作为空间的主图案，天花与地面相呼应，寓意着企业对每位员工平等、稳步发展的理念，并以此开发出相应的硬装元素及软装造型。开放式水吧台，雅座洽谈区，产品展示区，多功能休息区，在这里增加了人们相遇的机会，促进了广泛的交流，有着积极深远的意义。

This scenario is a retrofit design of an old factory office. In order to preserve the characteristics of the original factory area and integrate it with the atmosphere of other floor office areas, the LFOT industrial style was selected. The design is intended to break the dull, old-fashioned, intrinsic office model and create a passionate and distinctive office environment. Stepping into the sensor door, the designer skillfully uses the irregular geometric shape, and then paints the walls with bright yellow and fruit green to make the whole space full of fresh vitality. Taking the hexagon of the honeycomb as the main pattern of the space, the ceiling echoes the ground, which symbolizes the concept of equal and steady development of each employee and develops the corresponding hard-wearing elements and soft-packing styles. The open water bar, the elegant seating area, the product display area and the multi-functional rest area have increased the opportunities for people to meet and promoted extensive exchanges with positive and far-reaching significance.

桂林岩兰酒店
Guilin Yanlan Hotel

岩兰酒店致力于为白领精英旅游者提供更为周到的旅游咨询、旅游代办服务，让旅游者的出行拥有极高的品质。岩兰强调旅游住宿品质，希望为居住者提供更有氛围、更加轻松、更加舒适的旅游住宿体验。与咖啡吧台合二为一的酒店前台，不定时举办读书会、旅游分享会及新品发布会的大堂书吧，拥有极佳景观的露台餐厅都提供了与众不同的酒店体验。

The Yanlan Hotel is committed to providing white-collar elite travellers with more thoughtful travel advice and travel agency services, so to ensure the travellers a high quality of travel. Yanlan Hotel emphasizes the quality of travel accommodation and hopes to provide a more cozy, more relaxing and more comfortable travel experience for the elite travelers. The front desk of the hotel, also serving as the coffee bar, occasionally holds book clubs, sharing sessions of travel and lobby tables for new product launches. The terrace restaurant with excellent views provides a unique hotel experience for the travelers.

项目名称：桂林岩兰酒店
项目设计：广东星艺装饰集团
项目地址：广西桂林
设计师：范建国、许舰

3 公共·工程实景作品
Public · Engineering Live-scene Works

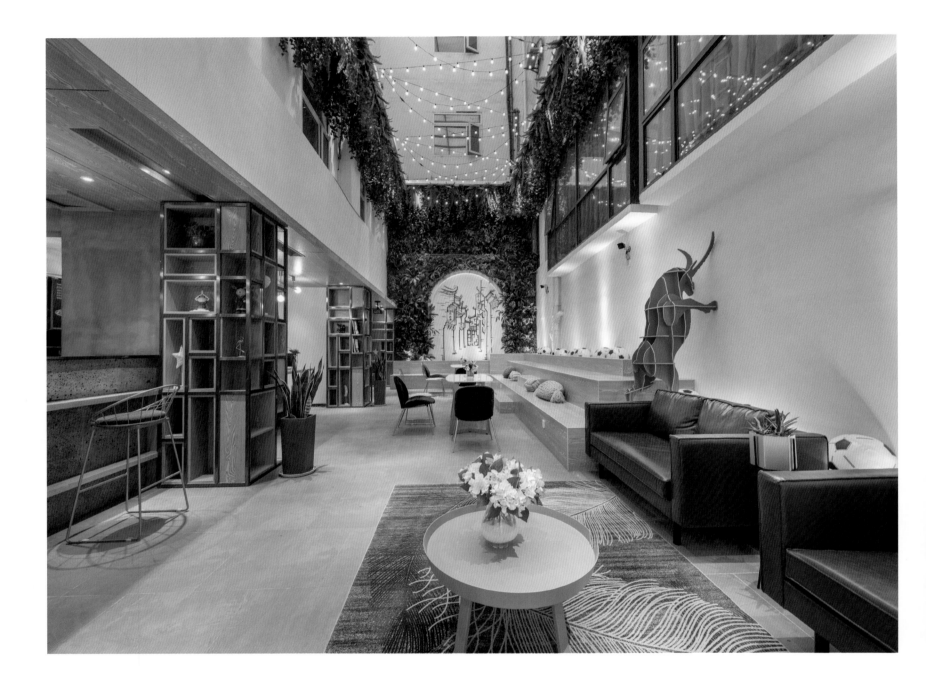

"白房子"创客中心之 418 CUCOLORIS

"White House" Makerspace at 418 Cucoloris

设计师希望让每个空间都能看到院子里的杨桃树，改造建筑大窗以增加景流量。二层露天成为一个公共的聚会场所，大门不上锁，告知新住民："任其自流，悠哉游哉。"

院子里的杨桃树，是每个房间的一景。一层原副体建筑，用工字钢重新做结构加固，把原有的楼板和圈梁用钢体支撑，让建筑、树和人融合在一起，外扩建筑形态依托钢架沿树体形态斜面而走，以树养形，以木性养心性。

In order to make each space see the carambola trees in the yard, the designer thus has transformed the large windows to expand the visions. The second floor has become a public meeting place, and the door is unlocked, telling the new residents, "Let it flow, and enjoy."

Each room can appreciate the view of carambola tree in the yard. A layer of the original auxiliary structure is re-reinforced with I-beams, and the original floor and ring beams are supported by steel bodies, so that the buildings, trees and people are merged together. The external expansion structure relies on the steel frame along the slope of the tree shape, and the tree is thus used to raise the shape and to cultivate the temperament with wood.

项目名称："白房子"创客中心之 418 CUCOLORIS
项目设计：广东星艺装饰集团
项目地址：广东东莞
设计师：陈文辉、洪易娜

瓶子酒吧
Bottle Bar

本案位于南京的鼓楼区大石桥羲和商业中心，中心集成商业办公、餐饮娱乐等项目，便利的交通、繁华的街道迎来了咖啡等休闲餐饮商业大咖。瓶子餐吧定位为以货真价实的洋酒及酒文化为主题的清吧，配合简单小吃、四川的串串，满屋的实木物件，舒适的真皮座椅，在柔和唯美的灯光下，享受舒适，享受生活。

This scenario is located in the Dashiqiao Yihe Commercial Center in Gulou District, Nanjing. The center integrates commercial office, catering and entertainment projects. The convenient transportation and bustling streets ushered in renowned coffee bars and other casual dining business. The Bottle Bar is positioned as a lounge bar with the theme of imported wine and wine culture. With simple snacks, Sichuan skewers, roomful wood items, comfortable leather seats, people could enjoy the comfort and life under the soft and beautiful lighting.

项目名称：瓶子酒吧
项目设计：广东星艺装饰集团
项目地址：江苏南京
设计师：李佩耀

点心时代
Snack Times

本案设计理念从蒸笼元素和大自然森林主题相结合出发，试图将大自然森林的氛围带入 300 多平方米的餐厅环境中来。设计中庞大的蒸笼元素造型，大小不一的组合，作为天花的唯一装饰出现在空间里，与白桦树结合混凝土柱体的默契融合，也是整个空间的视觉焦点。大面积的水泥地面、墙面、丹麦经典的 GUBI 金属吊灯，运用白桦树作为屏风，隐约地分割成了餐厅的几个区域，伴随着大自然最重要的颜色——草木绿色的窗帘、卡座皮革、植物作为空间点缀色，让空间如大自然般清新起来了。

The design concept of this scenario is based on the combination of steamer elements and natural forest themes, trying to bring the atmosphere of the natural forest into the restaurant environment of more than 300 square meters. The huge steamer elements in the design, the combination of different sizes, appear as the only decoration of the ceiling in the space, and the tacit fusion with the birch tree combined with the concrete cylinder is also the visual focus of the whole space. Large-scale concrete floors, walls, Danish classic GUBI metal chandeliers, using birch trees as screens, looming into several areas of the restaurant, accompanied by curtains, decks leather and plants in green, the most important color of nature to decorate the space, make the space as fresh as nature.

项目名称：点心时代
项目设计：广东星艺装饰集团
项目地址：江苏南京
设计师：黄剑

东方极韵
Oriental Charm

本案提倡自然简洁和理性，空间设计比例均匀、形式新颖，内部结构严密紧凑，空间穿插有序，围护体各界要素的虚实构成非常明显。另外，通过虚实互换的空间形象，取得局部与整个空间的和谐。

This scenario advocates natural simplicity and rationality, the spatial design ratio is uniform, the form is novel, the internal structure is tight and compact, the space is interspersed and orderly, and the virtual and real components of the various elements of the enclosure are very obvious. In addition, through the spatial image of the virtual and real interchange, the harmony between the local and the entire space is achieved.

项目名称：东方极韵
项目设计：广东星艺装饰集团
项目地址：天津市
设计师：董亮军

李白啤酒馆
Let's Beer

Let's Beer 是客户自创的品牌，是天津的第一家旗舰店。设计师结合品牌、天津本土特色以及消费人群等特点，给项目重新定位。随处可见的铁艺自选啤酒架，配合旧木板搓木蜡油；吧台摒弃了常见的高脚凳配合高吧台的做法，改为舒适的沙发以及低矮的吧台。二层围合空间的设计，让所有位置都能够自然地看到吧台正后方的投影屏幕；照明的设计考虑到空间的整体性、延伸性、引导性以及私密性。三层空间以包场聚会为主要功能，风格也延续了一、二层的设计理念。

Let's Beer is a customer-created brand and the first flagship store in Tianjin. The designer repositioned the project with the characteristics of the brand, local characteristics of Tianjin and the consumer group. The wrought iron beer racks are ubiquitous with the old wooden eucalyptus wax oil. The bar abandons the common high stools with the high bar, and changes to a comfortable sofa and a low bar. The design of the enclosed space on the second story enables all locations to naturally see the projection screen directly behind the bar; the design of the lighting takes into account the integrity, extensibility, instruction and privacy of the space. The three-story space is mainly composed of private parties, and the style continues the design concept of the first and second floors.

项目名称：李白啤酒馆
项目设计：广东星艺装饰集团
项目地址：天津市
设计师：马呈龙

芃泰科技
Pengtai Technology

本案位于天津市南开区金融中心。业主颇具文人气息,特别喜好现代干净整洁的办公环境。于是,设计师在与业主沟通后,把本案定位为现代简约风格,采用浅色的大理石、白色的护墙板,将空间打造成清新简约风格。

This scenario is located in the financial center of Nankai District, Tianjin. The owners are very literate and particularly fond of the modern clean and tidy office environment. Therefore, after communicating with the owner, the designer positioned the design as a modern minimalist style, using light-colored marble and white siding board to create a fresh and simple style.

项目名称:芃泰科技
项目设计:广东星艺装饰集团
项目地址:天津南开
设计师:李元青

天佑销售中心
Tianyou Sales Center

售楼处是形象重要的展示场所，也是一个独特广告载体。作为直接影响客户第一视觉的室内空间设计，更要体现楼盘所具有的特色及形象。良好的设计包装能给客户留下深刻的印象，更能第一时间争取客户，从而增强客户的购买欲。

The sales office is an important display place for the image and a unique advertising carrier. As the interior space design that directly affects the customer's first vision, it must reflect the characteristics and vivid image of the real estate. Good design packaging can impress customers, and can win customers in the first time, thus enhancing customers' desire to purchase.

项目名称：天佑销售中心
项目设计：广东星艺装饰集团
项目地址：云南红河州
设计师：钟华东

对于色调，设计师在设计时选用色调鲜艳的颜色，以求吸引客户的关注。对购房者而言，鲜艳的颜色具有亲和力，给人以朝气蓬勃的感觉，也能彰显其特点与性格。室内空间应该给每个客户营造家的温暖，因此，室内氛围很重要。在室内灯光设计上，不拘泥于形式，运用各种不同的灯光色彩使室内的气氛更加活跃、明快。

For color tones, designers use bright colors to attract customers' attention. For buyers, the bright colors endow with affinity, giving them a sense of vigor and vitality, as well as their characteristics and character.

The interior space should create a warm area for each customer, so the indoor atmosphere is important. In the interior lighting design, the style of the room is more active and bright, without formality.

室内空间适当地引入一些动态的元素，可以活跃室内气氛，对销售人员与客户的交流更具有推动性。售楼处是个功能多元化的地方，除了展示企业实力、文化、品牌之外还要体现以人为本、以客为尊的价值观，打造一个放松、舒适的环境，让客户感觉轻松、自在，签单率自然也会提高。

The indoor space appropriately introduces some dynamic elements, which can activate the indoor space atmosphere and promote the communication between the sales staff and the customers. The sales office is a function-diversified place. In addition to demonstrating the strength, culture and brand of the company, it also reflects the values of people-oriented and customer-oriented, creating a relaxed and comfortable environment, making customers feel relaxed and comfortable, and the rate of signing will naturally improve.

贺州星艺业务部
Hezhou Xingyi Business Department

本案整体空间让人眼前一亮，以灰白色为主色调，配以黑色、木色的点缀，冷静却不单调，直行线条贯穿整个办公空间，细腻而有层次，虚实量体组合的空间具有深度与灵韵，凸显出一种人性的温情和优雅，以绿色自然的设计理念来体现空间的视觉美感。

The overall space of this scenario can stun the customers, with gray-white as the main color, mixing with black and wood accents, calm but not monotonous. Straight lines run through the entire office space, exquisite and layered. The space of the combination of virtual and real body reveals depth and aura, highlighting a kind of human warmth and elegance, reflecting the visual beauty of space with the green and natural design concept.

项目名称：贺州星艺业务部
项目设计：广东星艺装饰集团
项目地址：广西贺州
设计师：余欢

4

公共·方案设计作品
Public·Scenario Design Works

创举办公室
Pioneering Office

作为一个集产品设计、生产与销售为一体的年轻团队，需要一个怎样的办公空间？
作为同样年轻的室内设计团队，应该怎样将空间赋予灵感？

As a young team that integrates product design, production and sales, what kind of office space do they need?
As an equally young interior design team, how to inspire space?

项目名称：创举办公室
项目设计：广东星艺装饰集团
项目地址：广东广州
设计师：谭立予

与他们的产品设计一样，既实用，又时尚；

与他们的产品设计一样，尊重材料，把各种材料放到正确的位置；

与他们的产品设计一样，找到设计与人、空间与人最本质的问题：比如除了合理地安排好每个空间功能，是否能让空间成为场所具有某种"场所精神"？

将上面几点都做好，看似简单，实则不易。

最后，白色钢柱，向密斯•凡德罗致敬！

As well as their product design, the design should be both practical and stylish;
As well as their product design, the design should make full use of the materials and put the materials in the right place;
As well as their product design, the design should find the most essential issues of design and people, space and people, for example, in addition to rationally arranging each spatial function, is it possible to make space a place with a certain "place spirit"?
It seems to be simple to meet the above requirements, but it is not easy.
At last, white steel column, salute to Mies Vander Rohe.

4 公共·方案设计作品
Public · Scenario Design Works

至和大厦顶层会所
Top Club in Zhihe Building

本案位于城区一栋商务楼的顶层。业主从商多年，一直想要一个符合自己理想的集接待、会客、休闲、办公、休息为一体的独立空间。

This scenario is located on the top floor of a commercial building in the city. The owner has been in business for many years and has always wanted an independent space that integrates reception, meeting, leisure, office and rest.

项目名称：至和大厦顶层会所
项目设计：广东星艺装饰集团
项目地址：山西大同
设计师：周 鹏、冷 叶

因为是顶层，且有自己独立的空中花园，所以设计师对此案有明确的空间概念，让空间在闹市中具备世外桃源的感觉。空间要做到一步一景，每个独立空间都有一种传统文化的元素融入，有水乡有黛瓦有行云有宋瓷有山水。

中庭是设计的重点，中间的天井能赏月观星，星艺设计九大家文化的融入，天、地、人三宝的运用，加上日出东方的墙景，使得空间气场十足。

Because it is the top floor and has its own independent sky garden, the designer has a clear space concept for the design, which makes the space feel like a paradise in the downtown area. The space needs to be a scene on step. Each independent space has an element of traditional culture, with water and grey tiles as well as mountains and rivers.

The atrium is the focus of the design. The middle patio is a good site to appreciate the night view with the moon and stars. The integration of the nine cultures of the star art design, the use of the heavens, the earth and the people, together with the sunrise of the eastern wall, make the space full of aura.

4 公共·方案设计作品
Public · Scenario Design Works

稻之源料理

Daozhiyuan Cuisine

自然节制，返璞归真。

本案设计核心在于打造一个自然、放松的环境，让人身处其中，视觉、听觉、嗅觉及触觉都能够体验到舒适与放松。整个环境用简洁的方式还原空间本质，通过自然的材料来打造纯净的空间。

Natural moderation, return to the truth.
The core design of this scenario is to create a natural and relaxed environment where people can experience comfort and relaxation in sight, hearing, smell and touch. The entire environment restores the essence of space in a concise way, creating a pure space through natural materials.

项目名称：稻之源料理
项目设计：广东星艺装饰集团
项目地址：广西南宁
设计师：凌立成

4 公共·方案设计作品
Public · Scenario Design Works

凤凰空间艺术馆

Phoenix Space Museum of Art

本案是一家以艺术拍卖和自由交易为主题的美术馆。业主多年从事艺术品收藏、鉴赏工作，有一定的社会知名度和美誉度。在设计前，设计师对工地进行了详细的勘察和对比，最终确定以"孔雀"为主题进行设计。从孔雀开屏到孔雀羽毛有序排列，主题元素贯穿整个空间，与精美的陶艺、高超的艺术相得益彰，美妙绝伦，让人沉浸在中国古典文化深厚的底蕴中，更加舒适地享受空间，享受生活。

This scenario is an art gallery with the theme of art auctions and free trade. The owner has been engaged in art collection and appreciation for many years and has a certain social reputation. Before the design, the designer conducted a detailed survey and comparison of the construction site, and finally decided to design with the theme of "peacock". From the opening of the peacock to the ordering of the peacock feathers, the theme elements run through the space, and the exquisite ceramic art and superb art complement each other. Immersed in the profound heritage of Chinese classical culture, it is wonderful to enjoy the space and enjoy the life.

项目名称：凤凰空间艺术馆
项目设计：广东星艺装饰集团
项目地址：广东广州
设计师：帅伯尤、曹彪

森山亭
Senshan Pavilion

森山亭——"森栅停"。

森：会意法造字，从林从木。本义：树木丛生繁密。"森"字为表现空间的结构形态，采用叠加组合方式构成。

栅：指用竹、木、铁条等做成的阻拦物。"结木为栅"中"栅"字表示围栏的意思。"栅"字为表现空间的功能形式，通过围栏方式进行分格。

Senshan Pavilion—Halt on seeing the railings.
Sen: The character is created by its shape and meaning, from the wood and the forest. Original meaning: dense trees. Sen is the structural form of the performance space, which is constructed by superposition combination.
Railing: refers to a block made of bamboo, wood, iron bars, etc. The word "railing" in the four-letter word "railings made of wood" has the meaning of the fence. The word "railing" is a functional form of performance space, and is divided by a fence.

项目名称：森山亭
项目设计：广东星艺装饰集团
项目地址：广东梅州
设计师：陈梓杰

停：意思是止住，终止不动，暂时不继续前进。亭，在古时候是供行人休息的地方。"亭者，停也。人所停集也。"《释名》园中之亭，应当是自然山水或村亭子镇路边之亭的"再现"。水乡山村，道旁多设亭，供行人歇脚，有半山亭、路亭、半江亭等，"停"字为表现空间的意义所在，人见亭而停，所以请停下匆忙的脚步，用心去感受乡村之美。

"江山无限景，都取一亭中。"这就是亭子的作用，就是把外界大空间的无限景色都吸收进来，为了使游览者从小空间进入大空间，也就是突破有限，进入无限。

Halt: refers to stopping, and not moving forward. "Ting"(pavilion)was a place for pedestrians to rest in ancient times, therefore, on seeing pavilion, the pedestrians halt. The pavilion in the "Shiming" should be the "reproduction" of the natural landscape or the pavilion on the roadside of the village. In the village abundant with water and mountain, there are many pavilions at the side of the road for pedestrians to rest, such as Banshan Pavilion, Lu Pavilion and Banjiang Pavilion. The word "halt" is the meaning of the performance space. People halt at the pavilion, so please halt your hasty steps and feel the beauty of the country with your heart.

"In a pavilion, people can have a panoramic view of the infinite scenery of landscape ." This is the role of the pavilion, which is used to absorb the infinite scenery of the large space outside, in order to allow visitors to enter the large space from a small space, that is, breaking into the infinity.

航帆酒店
Hangfan Hotel

本案是一个造型独特的精品酒店空间,以"突出重点,重视细节"为主题思想,表现出简洁、舒适、明亮的酒店空间。

This scenario is a unique boutique hotel space with the main idea of "highlighting key points and attention to detail", showing a simple, comfortable and bright hotel space.

项目名称:航帆酒店
项目设计:广东星艺装饰集团
项目地址:河北秦皇岛
设计师:夏文彬、崔 健

以现代风格为主，提取现代风格的线条精髓和块面元素相结合，贯穿在整个酒店空间，使之呈现出简洁大方、美观实用的室内空间。以温馨的暖色系为总体主导色调，以灰木纹瓷砖为主铺设地面，以暖色木饰面为主装饰墙面，吊顶以平顶为主，局部灯带，在控制成本的前提下，更加突出酒店的现代特征。

Based on modern style, the combination of the essence of modern style line and block surface elements runs through the entire hotel space, making it a simple, elegant and practical interior space. With warm colors as the overall dominant color, the gray wood tiles are mainly used to lay the ground, and the warm wood veneer is the main decorative wall. The ceiling is mainly flat with the local light strips. Within the budget of the cost, the modern character of the hotel is highlighted.

尚·灰
High-end Gray Space

设计师经与客户沟通后，在设计中特别注意几个要点，进门空间营造一个让人眼前一亮的原创设计，内部空间追求多样化，还单独设置了小孩游戏区和学习区。整体风格偏向现代时尚。

After communicating with the customer, the designer pays special attention to a few points in the design. The entrance space creates an original design which is stunning, and the interior space pursues diversification. Moreover, the children's play area and study area are separately set. The overall style is biased towards modern fashion.

项目名称：尚·灰
项目设计：广东星艺装饰集团
项目地址：福建 龙岩
设计师：邓文华

诚信保险代理呼和浩特市总部
Credit Insurance Agency Headquarters in Hohhot

本案位于写字楼北侧，方位属水，因此在设计中引入水的元素。前台为岛屿，地面白色和深灰色的地砖切割弧形，整体空间流动自然，泾渭分明。主体背景也使用了内凹的设计，表现出环抱的感觉。地、墙、顶的元素融合更让空间具有不拘一格的特色，象征着天、地、人结合的理念，步入其中，和谐融洽，悠然自得，富有启迪。

This scenario is located on the north side of the office building and is water-based, so the elements of water are introduced into the design. The front desk is an island, and the ground white and dark gray floor tiles are curved and the whole space flows naturally. The body background also uses a concave design that embodies the feeling of embracing. The integration of the ground, the wall and the roof makes the space have an eclectic character, and it symbolizes the concept of combining heaven, earth and people. Stepping into it, people could feel the harmonious, leisurely and enlightening atmosphere.

项目名称：诚信保险代理呼和浩特市总部
项目设计：广东星艺装饰集团
项目地址：内蒙古呼和浩特
设计师：刘 鑫

索兰国际
Solan

项目名称：索兰国际
项目设计：广东星艺装饰集团
项目地址：广东广州
设计师：江文锋

本案坐落在广州市白云区一座原素原创创意园内，是一个美容连锁品牌。在设计上，设计师运用黑白灰色调来体现现代都市生活节奏和时代的变化。前台区域设置了一个干景园林区，使空间有区域穿插的视觉感。公共办公区域设置了员工休息区，以榻榻米为设计抬高了地面，区分办公空间与休闲区域的空间，增加办公区域的娱乐性和个性。总经理办公室也是以黑、白、灰为基本色调，一幅个性的人物画装饰了移门，分隔了办公空间与休息室，做到动静结合但又互不干扰。

This scenario is located in an elemental original creative park in Baiyun District, Guangzhou, and is a beauty chain brand. In design, the designer uses black and white gray to reflect the modern urban pace of life and the changes of the times. A dry-view garden area is set in the reception area, giving the space a visual sense of interspersed areas. The public office area is equipped with a staff rest area, which is designed with tatami to raise the ground, distinguish between office space and leisure area, and to increase the entertainment and personality of the office area. The general manager's office also uses black, white and gray as the basic color. A personalized figure painting decorates the sliding door, which separates the office space from the lounge, so that it can be combined with each other without interference.

数控中心
Numerical Control Center

本案设计理念围绕"科幻"进行表达，在设计中大量使用了灰色调材质，定制水泥墙板，防静电自流平地面，通过蓝色的辅光源把整个空间的科技感衬托出来。天花运用了大面积黑色哑面钛金方通，让空间有更强烈的对比感。

The design concept of this scenario is expressed around "Science Fiction". In the design, a large number of gray tone materials are used, custom cement wall panels, anti-static self-leveling ground, and the blue space auxiliary light source to bring out the scientific and technological sense of the entire space. The ceiling uses a large area of black matte titanium square to make the space have a stronger contrast.

项目名称：数控中心
项目设计：广东星艺装饰集团
项目地址：广西贵港
设计师：蒋毅

同心美术馆
Tongxin Museum of Art

同心美术馆旨在提高当地公众文化修养，协助艺术教育，为艺术爱好者提供交流创作的有效平台，主要展示当地书画、漆器、蜡染等名家名作。

在设计创作过程中，设计师以"画屏"为主要应用元素，重新解构，使之可以成为一个具备灵活性的室内构件，并让"画屏"承载当地的艺术创作与文化，用留白手法使之贯穿整个公共展示空间，让空间更具有包容性。

The Tongxin Museum of art aims to improve the local public culture, assist art education, and provide an effective platform for art lovers to exchange ideas. It mainly displays famous local masterpieces such as paintings, lacquerware and batik.
In the process of design creation, the designer uses "painting screen" as the main application element and re-deconstructs it so that it can become a flexible indoor component, and enable the "painting screen" to carry the local artistic creation and culture. The white space runs through the entire public display space, making the space more inclusive.

项目名称：同心美术馆
项目设计：广东星艺装饰集团
项目地址：贵州毕节
设计师：冯立龙

一朝一春（伊豆原）
Yidouyuan Cuisine

本案设计体现了对料理文化的执着和对自然生活的追求，在安静和温馨的氛围中体现出对就餐环境的完美追求。本案设计与传统日式设计有所不同，采用的是现代的设计手法，门头采用大的色块形成简洁明了的视觉感受。结合镂空设计，在大的色块下形成一个"破"的格局，打破大面积所带来的"闷"和"压"的感觉。

The design of this scenario reflects the persistence of the cooking culture and the pursuit of natural life, reflecting the perfect pursuit of the dining environment in a quiet and warm atmosphere. The design of this scenario is different from the traditional Japanese design. It adopts the modern design method, and the door uses large color blocks to form a simple and clear visual experience. Combined with the "hollow out" design, a "broken" pattern is formed under the large color blocks, breaking the feeling of "boring" and "pressing" caused by large areas.

项目名称：一朝一春（伊豆原）
项目设计：广东星艺装饰集团
项目地址：浙江平湖
设计师：张 贵

在水一方
Waterside Boutique Hotel

阳朔在水一方酒店坐落于遇龙河景区最美地段，前望遇龙河，后览月亮山。

我们在城市的钢筋水泥中生活久了，越发向往陶渊明笔下的世外桃源。平淡无声的日子里，执一支素笔写下美丽的诗行。与爱人诉衷肠，和山水共清欢。

Yangshuo is located in the most beautiful section of the Yulong River Scenic Area, overlooking the Dragon River and the Moon Hill.
We have lived in the city's reinforced concrete for a long time, and we are increasingly looking forward to the paradise as Tao Yuanming described. Writing beautiful poems to share with lovers and take a delight in appreciating the nearby mountains and rivers.

项目名称：在水一方
项目设计：广东星艺装饰集团
项目地址：广西桂林
设计师：陆　勇

迈卡酒店
Maika Hotel

酒店以时尚、东方、休闲、温馨为设计理念，整体设计以新东方文化元素为主，并通过中西组合家具、陈设以及艺术品的巧妙装饰，呈现出静谧自然的时尚东方气质。

酒店整体用色和谐统一，辅以简洁明快的线条，以现代风格设计手法，结合用心调试的光源，让空间散发出时尚、宁静、优雅、温馨的东方气质。它既是一次东方文化呈现，也是对中式风格酒店的一次新的探索和诠释。

The hotel is designed with fashion, oriental, leisure and warmth. The overall design is based on the elements of New Oriental culture. Through the combination of Chinese and Western furniture, furnishings and art, it presents a quiet and natural oriental temperament.

The hotel's overall color is harmonious and unified, supplemented by simple and bright lines, with modern style design techniques, combined with the light source for debugging, let the space exude a fashionable, quiet, elegant and warm oriental temperament. It is both a representation of oriental culture and a new exploration and interpretation of Chinese-style hotels.

项目名称：迈卡酒店
项目设计：广东星艺装饰集团
项目地址：广东惠州
设计师：何焱生

聚和兴茶庄
Juhexing Tea House

"美"总在无限想象里,"韵"总在虚实相生间。恰到好处的留白不仅可以给人审美的享受,还能构造出空灵的韵味。本案设计中一面白墙,寥寥数笔,形成泼墨般的视觉效果,尽显含蓄端庄的东方韵味。自古以来,中国人深受"天圆地方"观念的影响,人们会将这一美学精粹呈现于建筑空间中,营造出风雅、开阔的空间意境。在墙面上做留白,可以凸显出强烈的立体感,空间形状的鲜明对比,于方寸之间自成天地。

"Beauty" is always in the imagination of infinite, while "charm" in the endless artistic conception. The right blank can not only give people an aesthetic enjoyment, but also construct an ethereal charm. In the design of this scenario, a white wall and a few pens form a visual effect of splashing ink, showing the subtle and dignified oriental charm. Since ancient times, the Chinese have been deeply influenced by the concept of "the place of heaven and earth". People will present this aesthetic essence in the architectural space, creating an elegant and open space. White space on the wall can highlight the strong three-dimensional sense, the sharp contrast of the space shape, between the square inch of the world.

项目名称:聚和兴茶庄
项目设计:广东星艺装饰集团
项目地址:云南昆明
设计师:熊卫星

素菜馆
Vegetarian Restaurant

项目名称：素菜馆
项目设计：广东星艺装饰集团
项目地址：广西贺州
设计师：黄林妮

本案素菜馆多以轻松、舒畅为空间主导气氛，通过洁净的装修、淡雅的色彩，结合艺术性陈设，以及光影的变换，营造出一个新的具备"色、香、味、触"元素的餐饮环境。主要的界面在于吊顶、墙面和隔断的设计。由于本餐厅的层高较高，利用这个优势，将吊顶与柱子相结合，采用流线型设计高低起伏不一的流线形态，使整个空间产生节奏感、韵律感。此外，通过色调来营造环境氛围，以自然木质、白色和灰色协调统一，运用到空间中。光在色调上也起到相当重要的作用，不仅能够潜移默化地影响食客的味觉，也能更好地烘托环境。另一面意在传达素食理念，让人类更加重视健康饮食，重视人与自然的关系。人类要尊重自然、重视自然、与自然和谐相处。

In this scenario, the vegetarian restaurant is more relaxed and comfortable as the space-oriented atmosphere. Through clean decoration, elegant colors, combined with artistic furnishings, and the transformation of light and shadow, a new dining environment with the elements of "color, fragrance, taste and touch" is created. The main interface is the design of the ceiling, wall and compartments. Due to the high floor height of the restaurant, this advantage is combined with the ceiling and the column, and the streamlined design has a streamlined shape with different undulations, which makes the whole space rhythmic. In addition, the atmosphere is created by the color tone, and the natural wood, white and gray are coordinated and applied to the space. Light also plays a fundamental role in the color tone, not only can subtly affect the taste of the diners, but also better to improve the environment. The other side is intended to convey the concept of vegetarianism, so to arouse human beings' attention to healthy eating and to the relationship between man and nature. Human beings must respect nature, value nature, and live in harmony with nature.

耕艺种德

设计幸福